Expanding the
UNITED
STATES

Contents

Chapter 1

Moving West

"We are heading west!" This is what many people said in the mid-1800s when they planned to go west. And they did go west. They rode horses, walked, and traveled by covered wagon. Men, women, and children of all ages left their homes with one dream. They hoped to find land.

Gold was found in California in 1848. News of the find spread quickly. Soon, many people in the east were excited about stories of gold.

TALK ABOUT IT

If you told ten friends about the discovery of gold in California and each friend told ten more friends, how many people would know in all?

3

Life in the east was hard for people. The soil was not good for farming. Crops did not grow well. Many people could not find jobs. Many people were poor. News of gold in California gave these people new hope.

As soil in the east became poor, settlers looked to the rich soil of the far west.

Chapter 2

The Pioneers

Many people left their homes in the east for a better life in the west. These people were called pioneers. *Pioneer* means someone who leads the way for others.

This picture shows a wagon train traveling through the mountains during the late 1800s.

There were no trains, planes, or cars in the early 1800s. People traveled by horse and wagon.

Pioneers had a long way to go. There were more than 3,000 miles between the East Coast and California.

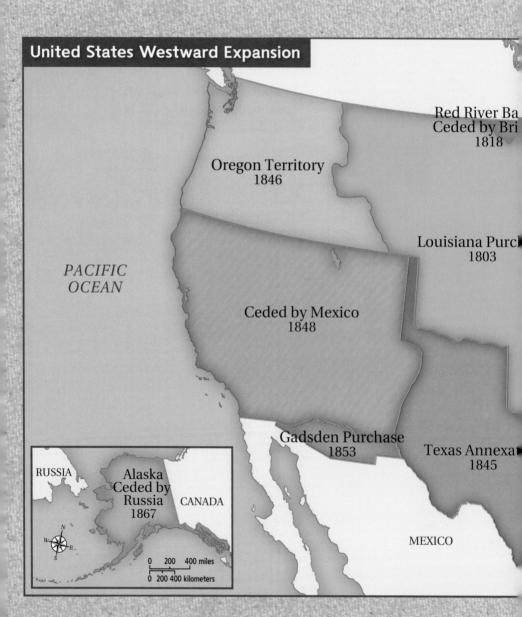

United States Westward Expansion

Red River Ba
Ceded by Bri
1818

Oregon Territory
1846

Louisiana Purc
1803

PACIFIC
OCEAN

Ceded by Mexico
1848

Gadsden Purchase
1853

Texas Annexa
1845

RUSSIA

Alaska
Ceded by
Russia
1867

CANADA

MEXICO

0   200   400 miles

0   200 400 kilometers

The trip was slow and hard. The roads between the east and California were not paved. In some places, there were no roads at all. Covered wagons could travel only 2 miles per hour.

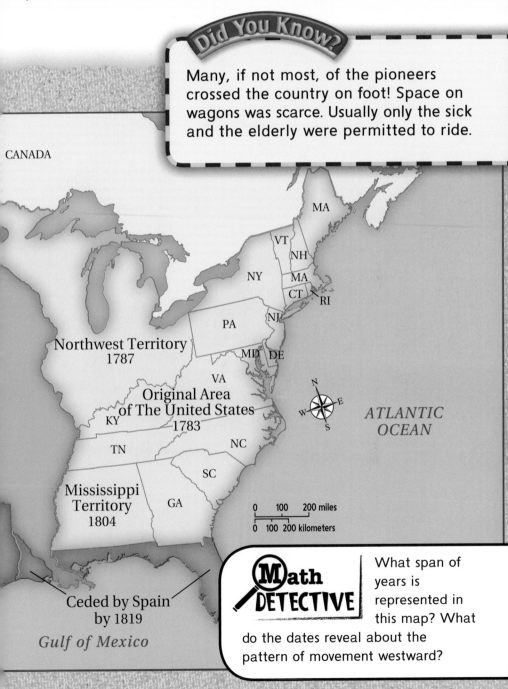

Did You Know?

Many, if not most, of the pioneers crossed the country on foot! Space on wagons was scarce. Usually only the sick and the elderly were permitted to ride.

CANADA

MA

VT

NH

NY

MA

CT

RI

NJ

PA

Northwest Territory
1787

MD DE

VA

Original Area
of The United States
1783

KY

N

E

W

S

ATLANTIC
OCEAN

TN

NC

SC

Mississippi
Territory
1804

GA

0   100   200 miles

0   100 200 kilometers

Math DETECTIVE

What span of years is represented in this map? What do the dates reveal about the pattern of movement westward?

Ceded by Spain
by 1819

Gulf of Mexico

Many people became sick on the trip. There was little food and water along the trail. Many people died before they got to California.

About 20,000 people died on the trail during the years of the westward expansion. For every 1 mile of trail, about 9 people died. Some of the graves can still be seen today.

A gravemarker shows where one person died during her journey west. People risked their lives in the hopes they could get their own land. During the 1840s, married people who reached Oregon were given 640 acres of land.

Chapter 3

Settling Along the Way

Some people did not make it to California. Some people decided to make new homes along the way. They farmed the land. They hunted. They traded goods. As more and more families settled the land, small villages formed.

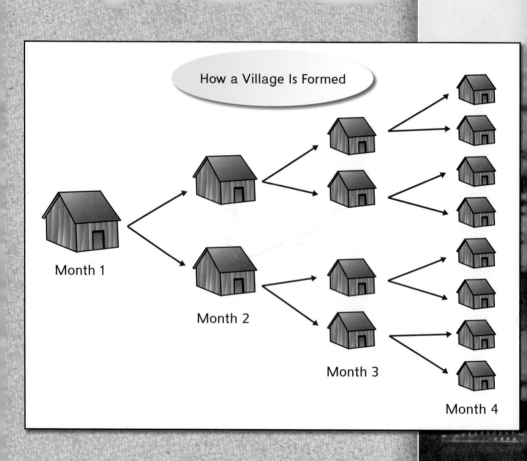

How a Village Is Formed

Month 1

Month 2

Month 3

Month 4

Villages grew. They became towns. But that was nothing compared to what happened in California.

The town of Helena, MT, in 1865. This capital city started out as a pioneer town.

It could take up to 6 months to cross the country. However, many people finally made it to California. In fact, about 300,000 people came to California during the California Gold Rush.

Gold mine towns formed very quickly. More than 80,000 people settled in California between 1848 and 1849.

Gold-Seekers Arriving in California		
Year	1848	1849
Number of Gold-Seekers	6,500	90,000

Chapter 4

California Living

Some people really did get rich from finding gold. It took a little digging and being in the right place.

Gold pan filled with gold.

Portsmouth Square in San Francisco, CA, during the 1850s.

For people who did not find gold, there were many other ways to make money in the west. Some people opened restaurants. Others opened laundries.

Women did not usually work in the east. In California, women did work. Women were needed for cooking and cleaning. Businesses grew across the state.

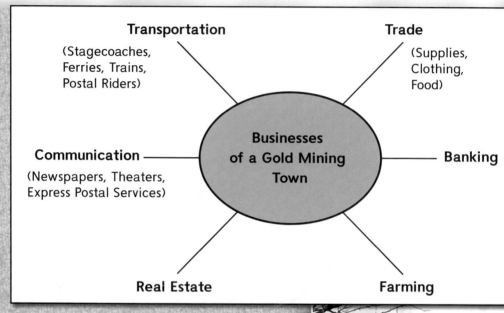

Transportation
(Stagecoaches, Ferries, Trains, Postal Riders)

Trade
(Supplies, Clothing, Food)

Communication
(Newspapers, Theaters, Express Postal Services)

Businesses of a Gold Mining Town

Banking

Real Estate

Farming

Did You Know?

A California settler could earn $25 for every $1 he or she made in the east.

One of the most famous people to come out of the Gold Rush was Levi Strauss. He made pants from canvas. He finally started his own company in 1853. In fact, you might be wearing his invention right now. Today, we call them blue jeans.

A modern pair of blue jeans.

With so many people moving to California, there was not enough food and supplies for everyone. When the need, or demand, for items is more than the supply of those items, prices go up.

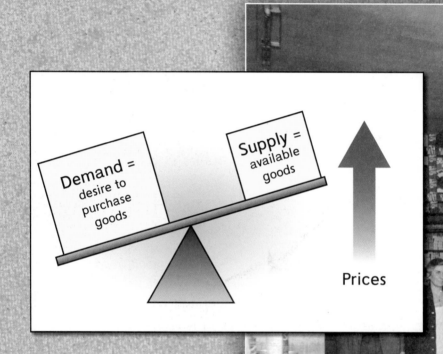

Demand = desire to purchase goods

Supply = available goods

Prices

TALK ABOUT IT

During this time, $25 in the east would buy a lot more than $25 in California. Why is this?

16

In the mid-1800s, some supplies in the west cost a lot more than what they cost in the east. Compare the prices.

Price Comparisons		
Item	Price in the East	Price in the West
Shirt	$4	$36
Boots	$2	$16
1 Dozen Eggs	20¢	$2

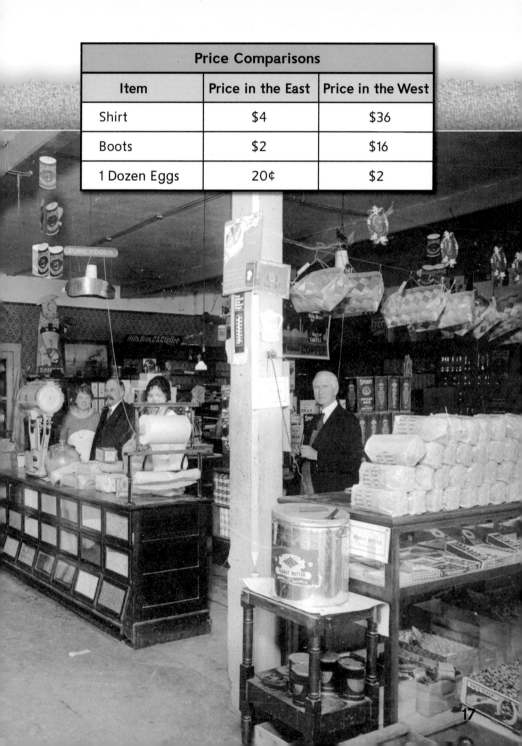

Most people thought there was plenty of gold. But as more and more people arrived in California, there was not enough gold for everyone. In 1849, about 6,000 people came to California. Thousands more followed. By 1850, there were more miners than gold.

1849	
Gold Supply	Miners
⃝⃝⃝⃝⃝⃝⃝⃝⃝ ⃝⃝⃝⃝⃝⃝⃝⃝⃝ ⃝⃝⃝⃝⃝⃝⃝⃝⃝ ⃝⃝⃝⃝⃝⃝⃝⃝⃝ ⃝⃝⃝⃝⃝⃝⃝⃝⃝ ⃝⃝⃝⃝⃝⃝⃝⃝⃝	☖☖☖☖☖☖☖☖☖☖ ☖☖☖☖☖☖☖☖☖☖

1850	
Gold Supply	Miners
⃝⃝⃝⃝⃝⃝⃝⃝⃝ ⃝⃝⃝⃝⃝⃝⃝⃝⃝	☖☖☖☖☖☖☖☖☖☖ ☖☖☖☖☖☖☖☖☖☖ ☖☖☖☖☖☖☖☖☖☖ ☖☖☖☖☖☖☖☖☖☖ ☖☖☖☖☖☖☖☖☖☖ ☖☖☖☖☖☖☖☖☖☖

TALK ABOUT IT

Gold supplies were on their way down. How would this affect the miners' wages?

Miners earned up to three times less money in 1850 than in 1849. Some miners gave up and went home to the east. Others stayed to find more gold. Those who had started businesses tried to survive. But when there was little gold left, tempers showed. Settlers fought each other for gold.

Miner Wages During the California Gold Rush			
	1 Day	1 Week	1 Month
1849	$42	$294 ($42 x 7)	$1,176 ($294 x 4)
1850	$14	$98 ($14 x 7)	$392 ($98 x 4)

GOLD MINES OF CALIFORNIA!!

W. R. ANDREWS,

Having just returned from California, after having spent several months in the mines and mountains of that interesting country, will deliver a

LECTURE,

on 1849,

at   day of

upon the GOLD MINES OF CALIFORNIA, and give a narrative of his journey to California, a description of the route and of the different places on the route to San Francisco, interesting incidents, &c. together with a full and accurate description of San Francisco, San Rafel, Benetia, Pacific New York, Mission Doloros, San Jose, Sutters' Fort, Suttersville, Stockton, and many other places in California, of the bugs, animals, vegetable productions, laws, the

Life in the west was especially hard for Native Americans. Many were forced to leave their homes. They lost their land and their way of life. Many Native Americans moved or died.

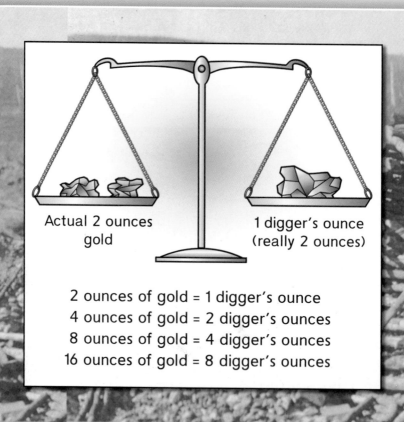

Actual 2 ounces gold

1 digger's ounce (really 2 ounces)

2 ounces of gold = 1 digger's ounce
4 ounces of gold = 2 digger's ounces
8 ounces of gold = 4 digger's ounces
16 ounces of gold = 8 digger's ounces

Did You Know?

Miners cheated Native Americans by weighing their gold with a "digger's ounce." A digger's ounce was actually a 2-ounce weight. The settlers falsely claimed that it was one ounce so that they could pay the Native Americans less for their gold.

Other groups of people suffered as well. People had come to California from Mexico, South America, Europe, and China. The United States citizens did not want to share the gold. So the government made every person who was not a citizen pay $20 in taxes each month.

Chapter 5

A New Country

The California Gold Rush changed the west forever. Many beautiful landscapes and clean waterways were gone. Cities appeared. About 30 new houses were finished every day.

As a result of westward expansion, towns and cities formed across the entire country. The East Coast and the West Coast were connected. California became an official state in 1850.

Success in the west came at a high price. But the early settlers led the way for others to live in the west. As more people moved west, the United States grew to its current size.

The first railroad was finished in 1869. The last spike put in place was a gold spike known as the "Golden Spike."

1. Look at page 7. A pioneer family travels for 30 hours at maximum speed in a covered wagon. Their neighbor walks 20 hours at a pace of 3 miles per hour. Which group traveled the greater distance? Explain your answer. [Chapter 2]

2. Look at page 9. A tourist bikes 24 miles along the pioneers' trail to California. How many graves can he or she expect to pass? Explain your reasoning. [Chapter 2]

3. Look at page 14. A pioneer earned $8 each week in the east. Predict the amount of money that this pioneer will earn each week in California. [Chapter 4]

4. Look at page 20. A pioneer measures out 34 digger's ounces of gold. How much gold is that in actual ounces? [Chapter 4]

5. Look at page 21. How much tax would a non-citizen pay in 6 months of mining? [Chapter 4]

6. Look at page 22. On average, how many houses were completed every month in California in the 1850s? Explain how you solved this problem. [Chapter 5]

Real-World
Problem Solving Library
MATHEMATICS

ISBN: 978-0-02-100909-1
MHID: 0-02-100909-0

9 0 0 0

Mc
Graw
Hill
Education

9 780021 009091

AMERICANS
ON THE
MOVE

About the Cover

When might the picture on the cover have been taken? Explain your reasoning.

Cover Credit: Steve Allen/Brand X Pictures

Send all inquiries to:
McGraw-Hill Education
8787 Orion Place
Columbus, OH 43240-4027

ISBN: 978-0-02-100906-0
MHID: 0-02-100906-6

Printed in the United States of America.

8 9 10 DOC 20 19 18 17 16 15 14